Cut the Cord

TV without Cable or Satellite

Guide to Free Over the Air Television and Internet Streaming

THOMAS HYSLIP

Copyright © 2017 Thomas Hyslip

All rights reserved.

ISBN: 1975859189
ISBN-13: 978-1975859183

So you have heard all the news about cutting the cord, and read how much money you can save by ditching cable. While it is true that you can save money by getting rid of cable or satellite television, but there are many things to consider when you make the decision to cut the cord. Therefore, I have written this book as a guide to help you make the correct decisions and obtain the best possible setup at the lowest cost.

As you will read, there are many different options and depending on your location and viewing pleasures, every situation is unique. Therefore, I have broken the book into 3 sections.

Sections 1 and 2 describe the 2 options for receiving television without cable or satellite, over the air antenna and live Internet streaming. Section 3 provides recommended setups for different channel and viewing preferences.

Section 1 (OTA) provides detailed information on antennas, signal strength, amplifiers, and how to use an over the air antenna for multiple televisions throughout your home. There is also information on different DVRs available for use with OTA signals and their functionality.

Section 2 (Internet) lists the different live and replay options, and provides a comparison of the different services. Section 2 also provides a review of the different devices for streaming television via the Internet.

Section 3 of the book is a reference for the recommended setup for different viewing options, such as free, sports, news, and movies.

There is also an appendix with tables that show you what channels are available via which streaming service. As well as which local channels are available and which streaming devices support which services.

Good luck, enjoy the book, and thank you!

Table of Contents

Section 1 6
Chapter 1: Over the Air (OTA) Television 6
Chapter 2: Antennas 8
Chapter 3: Television Setup 13
Chapter 4: Tivo DVR 20
Chapter 5: Channel Master DVR 21
Chapter 6: Tablo DVR 22
Chapter 7: HDHomeRun 24
Chapter 8: AirTV Player 26
Section 2 26
Chapter 9: Sling Television 27
Chapter 10: PlayStation Vue 35
Chapter 11: Hulu Live TV 40
Chapter 12: YouTube TV 43
Chapter 13: Direct TV Now 46
Chapter 14: Fubo TV 54
Chapter 15: Pluto TV 57
Chapter 16: Netflix 60
Chapter 17: Hulu 61
Chapter 18: Amazon Prime 61
Chapter 19: Vudu 61
Chapter 20: Tubi TV 62
Chapter 21: Crackle 62
Chapter 22: Ameba TV 63
Chapter 23: Roku 63
Chapter 24: Chromecast 64

- **Chapter 25: Apple TV** 65
- **Chapter 26: Amazon Fire TV** 66

Section 3 66
- **Chapter 27: Free Television** 66
- **Chapter 28: Sports** 67
- **Chapter 29: Movies** 67
- **Chapter 30: News** 68

Conclusion 68

Appendix 69
- **Streaming Device and Live Television Services** 69
- **Live Television Channels available on Streaming Services** 70
- **Live Local Channels** 78

Section 1

Chapter 1: Over the Air (OTA) Television

The best thing about over the air (OTA) television is that it is absolutely free! That's right it doesn't cost a dime, and even more amazing, the quality is often better than cable or satellite signals. Hard to believe isn't it? But, the OTA signal is not compressed like the cable and satellite signal. Therefore, it is often a better picture quality. For anyone that has not viewed OTA television since the conversion to digital frequencies from analog in 2009, you are truly in for a real surprise. So lets get started.

The first consideration when deciding if you will use an antenna for OTA broadcast television is where you live. If you live in or near a major metropolitan area, then you will be able to receive multiple OTA channels. But, if you live in the country you may only receive a few channels, or maybe none at all. Therefore, the first step is to check your local area for OTA broadcasts. I prefer to use the Antenna Web website at https://www.antennaweb.org/Address. But there are other websites as well, https://otadtv.com/, and https://www.tablo.com/tools/ are two more. Here you enter your home address and you receive a listing of all TV towers within range, the distance to the towers, the direction (azimuth) to the tower, and the recommended minimum antenna type to receive the channel. We will discuss antenna types and recommendations later. With this information, you can decide if it is worthwhile for you to purchase an antenna and test your reception.

I live outside Raleigh, NC, and Figure 1 shows the listing of channels I may receive from Antenna Web. Keep in mind, most OTA channels also broadcast sub-channels. So for example, in Raleigh, WTVD ABC 11, has 11.1, 11.2, and 11.3. Figure 2 shows the total possible channels based on my address at otadtv.com. But if you enter my home town of Wellsville, NY 14895, you will see there are no OTA channels available. While New York, NY 10001 can receive 80 OTA channels.

Cut the Cord: TV without Cable or Satellite

Stations		Antenna
WTVD-DT 11.1 ABC RF Channel: 11 19 miles at 197°	abc	Yellow
WRAL-DT 5.1 NBC RF Channel: 48 18 miles at 196°	NBC	Yellow
WRAZ-DT 50.1 FOX RF Channel: 49 18 miles at 196°	FOX	Yellow
WRDC-DT 28.1 MNT RF Channel: 28 18 miles at 198°	myTV	Yellow
WNCN-DT 17.1 CBS RF Channel: 17 18 miles at 196°	CBS	Yellow
WLFL-DT 22.1 CW RF Channel: 27 18 miles at 196°	CW	Yellow
WUNC-DT 4.1 PBS RF Channel: 25 38 miles at 272°	PBS	Red
WRAY-DT 30.1 SAH RF Channel: 42 20 miles at 120°	TCT	Red
WAUG-LD 8.1 IND RF Channel: 8 12 miles at 225°	IND	Red
WRPX-DT 47.1 ION RF Channel: 15 20 miles at 65°	ion	Blue
WACN-LP REL RF Channel: 34 18 miles at 196°	DAYSTAR	Blue
WUVC-DT 40.1 UNI RF Channel: 38 40 miles at 233°		Violet
WWIW-LD 66.1 REL RF Channel: 45 14 miles at 226°	REL	Violet
WZGS-CD 44.1 IND RF Channel: 44 17 miles at 241°	IND	Violet

TV Ch	RF Ch	Callsign	Sub-Channels
47	15	WRPX-TV	47.1 ION 47.2 Qubo 47.3 ION Life 47.4 Infomercials 47.5 QVC 47.6 HSN
17	17	WNCN	17.1 CBS 17.2 Antenna TV 17.3 Justice Network
22	27	WLFL	22.1 CW 22.2 American Sports Network
5	48	WRAL-TV	5.1 NBC 5.2 Heroes & Icons
50	49	WRAZ	50.1 FOX 50.2 Me-TV
28	28	WRDC	28.1 MyN 28.2 Grit 28.3 Comet TV
11	11	WTVD	11.1 ABC 11.2 Live Well 11.3 Laff
34	34	WARZ-CD	34.1 Retro TV 34.2 WARZ 34.3 Test Pattern
8	8	WAUG-LD	8.1 Independent 8.2 The Family Channel 8.3 PBJ
66	45	WWIW-LD	66.1 Daystar
40	38	WUVC-DT	40.1 Univision 40.2 UniMas 40.3 Bounce TV 40.4 GetTV
4	25	WUNC-TV	4.1 PBS 4.2 PBS Kids 24/7 4.3 PBS Encore 4.4 North Carolina Channel
40	40	WTNC-LD	40.1 Univision 40.2 UniMas 40.3 Bounce TV 40.4 GetTV

Another important consideration with antenna reception is the height of the antenna and obstructions from buildings, trees, and terrain. Antenna Web asks if your antenna will be installed 30 feet above ground. The difference can be profound. In my location, I lose 2 towers and 2 channels if the antenna is installed below 30 feet. Thankfully, Raleigh, NC is relatively flat and terrain does not significantly affect OTA reception. So as you can see, the possible channels available varies greatly based on location and topography.

Chapter 2: Antennas

So, if the news was good and you are able to receive OTA channels at your home, the next step is to purchase an antenna. There are many considerations when choosing an antenna. They include

Indoors vs Outdoors

Single TV or Multiple TVs

Distance to broadcast towers

Direction to broadcast towers

Installation height

The first antenna I recommend anyone purchase is a cheap omnidirectional indoor antenna. This will allow you to quickly test your signal, and try OTA television before you spend more money on better antennas and equipment. Depending on your location and situation, this may be all you need to get OTA channels. I highly recommend Channel Master products, and they have a "FLATenna 35" for $10 at www.channelmaster.com.

FLATenna 35 (Black)

Channel Master FLATenna 35, omni directional antenna

To install this antenna, simply connect the coaxial cable to the TV tuner (coaxial) input on the television. Then mount the antenna on the wall or window. For best reception you want to "point" the antenna in the direction of the broadcast towers. Antenna Web provides a handy map with tower locations and direction, so it is easy to orient the antenna in the correct direction. If there are more than one antenna in your location, you have 2 options. But first you have to decide if more channels, or specific channels are more important to you. If you must have a specific channel, then orient the antenna in the direction of the

broadcast tower for that channel. Since this is an omni-direction antenna you will still receive other towers, but maybe not all towers. During this initial test of your first OTA experience, try different directions and see what different channels you receive. You may be surprised by the difference. If you do not receive many channels, or none at all, then you may need a bigger antenna, a preamplifier, or possibly an outdoor antenna. Putting the antenna inside is convenient, but you lose some signal due to the building and roof. We will cover all these different options later in the book.

One issue that most first time cord cutters will immediately notice the first time they watch OTA broadcast television is the lack of a "guide" option. But don't worry if you also had a panic attack when you tried to view the guide of upcoming shows, there are many options available. However, if you chose to simply connect an antenna directly to your TV, you will either have no guide, or only a listing of current shows.

Antenna Types

If you recall from the channel listing from Antenna Web, there were recommended antennas with color codes for each tower/channel. These codes are called the CEA certified antenna color codes, and there are 6 color codes.

Yellow = Small multi-directional. Good for use in large metro areas.

Green = Medium multi-directional. Slightly more powerful than yellow and good for cable runs of over 20 feet.

Light Green = Large multi-directional or small directional with pre-amp. The most powerful multi-directional antenna and fist directional antenna.

Red = Medium directional. Most popular outdoor antenna type.

Blue = Medium directional with pre-amp. Adding a pre-amp to a red antenna helps with weak reception to increase available channels.

Violet = Large directional with pre-amp. Largest outdoor antennas. Used in rural areas with weak signals for maximum possible reception.

The color codes are listed from smallest to largest antenna, which also equates to reception ability. These color codes were created for outdoor antennas, but you will also see them used with indoor antennas. Although the indoor antennas will have additional interference and the mileage range may not be accurate, the color codes still give you an idea of what to expect from the indoor antenna.

Now that you have tried the OTA television it is time to pick a permanent antenna. The Channel Master FLATenna 35 is coded, yellow/green/red, so it meets the standard for those three categories. If you were happy with the reception and number of channels, then you are all set and can continue to watch OTA broadcasts with the antenna.

But, if your location called for a larger antenna or your reception was poor, you will need to upgrade to a new antenna. The first decision you need to make is where will you install the antenna, inside or outdoors? Many people choose to install their antenna in the attic, while others will mount the antenna outdoors for the best reception possible. I personally installed my antenna in my attic, but eventually moved it outside for better reception. There are many conditions that affect your reception and it is usually a matter of trial and error to see what works best. Another consideration is the need for cabling from the antenna to the television(s). Obviously, a single TV is very simple to connect, but you many have to run coaxial cable from the antenna to the TV. I was able to use the existing coaxial in my house with a few modifications. I will provide detailed instructions on how to accomplish this later in the book.

For attic installations, I recommend one of the following antennas depending on the space available for install. The Channnel Master ULTRAenna 60 and EXTREMETenna 80 are great multidirectional antennas, but the EXTREMETenna 80 requires a lot of space. It should be noted these can also be installed outdoors and are rated as such.

Another great option for indoor installation is the ClearStream 2V UHF/VHF indoor/outdoor antenna. I used this antenna for a while but eventually changed to a directional antenna when I moved the antenna outdoors. I currently use the RCA compact Yagi HDTV VHF/UHF antenna (ANT751R). I am very happy with the reception and even though this is a directional antenna, I still receive many stations over a 90 degree span. To ensure I received the local, ABC, NBC, CBS, and FOX channels I oriented the antenna in the direction of those towers (they are all located in one area) but I still receive PBS, CW, and a few other channels spread across my metro area.

A good recommendation when deciding between a directional, or multi-directional antenna is: If your broadcast towers are within 90 degrees of each other than a directional antenna is recommended If the broadcast towers are more then 90 degrees apart but less than 180 degrees, then a multi-directional antenna is recommended (see figure below).

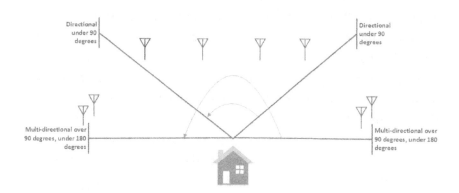

An important piece of information for anyone living with a homeowners association (HOA). The Telecommunications Act of 1996, section 207 specifically prohibits any HOA, condo association, landlord, state or local governments to "enact restrictions that impair the installation, maintenance, or use of antennas used to receive video programming." This is further codified in Title 47 CFR section 1400. The law also prevents the HOAs from requiring prior approval, professional installation, notification from a satellite provider that installation is necessary in a certain location, or any other requirement that would unnecessarily delay installation. So, you are covered by federal law to install your antenna wherever you receive the best reception if it is on your property (rentals included).

Now that you have tried the OTA television and picked an antenna, you have to decide how you will setup your OTA television for permanent use. I have provided numerous scenarios below, so pick the one that works best for you.

Chapter 3: Television Setup
Single Television

The easiest setup is a single television that will receive OTA broadcast channels and a small omni-directional antenna. This is great for small apartments and condos. If the FLATenna 35 I discussed in the previous section worked, and you were happy with the channels you received, then all you need to do is install the antenna in a good location. This is also true if you have more than one television and simply want to install a small antenna at each television location.

If you upgraded to a larger antenna for better reception, then you may need to install/connect coaxial cable between the television and the antenna, or use the existing coaxial cable in your home. See the section below about using your existing cable.

Multiple Televisions

If you are like me and have multiple televisions that require an antenna for OTA reception, but a small indoor antenna was not sufficient. Then you need to connect your large indoor or outdoor antenna to your televisions. There are some important factors to consider in this situation.

- How far is the antenna from the televisions?

- How many televisions do you have to connect?

- Do you have existing coaxial cable in the house?

The distance and number of televisions are important because of signal loss over the coaxial cable and signal loss when "splitting" the cable for additional televisions. I will not get into technical decibel loses, but a good rule of thumb is 1.5-6db loss per 100 feet of cable, and 3.5db loss for each "split." So, the more televisions and longer cable runs you have, the more signal lose you will incur. Therefore, it is important to plan for the losses and use amplified splitters, which is covered below.

A good way to overcome running new coaxial cable is to use any existing cable in your house. But, you must first understand what you have. For anyone with current or previous cable television, not satellite, then you will be able to use the cable as is. But, current or previous satellite customers will need to disconnect the powered single wire multiswitch (SWM) if present.

The SWM is connected between your satellite box and the dish. The coaxial cable will be plugged into it, and also needs to have power supplied through an outlet. If you have removed your satellite box and unplugged all the connections, then you already removed this device. Below is a picture of sample SWM power supply, and you can see the coaxial connections, as well as the power cable.

The actual SWM will be located somewhere at your house and you can look for it near the satellite dish or other telecommunication connections (telephone, cable, etc). Below is a picture of a sample SWM, but it may not look exactly like this. However, they will all look like a typical "splitter" with numerous coaxial cable connections. If you have cable television, you will most likely have a splitter that looks similar to this as well. From here, we will test the cable to see where each leads and label them.

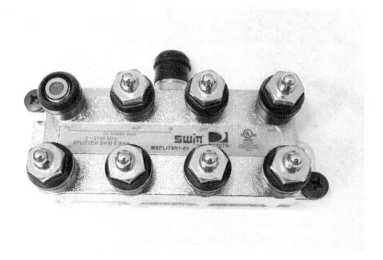

I recommend using the Klein Coax Explorer Plus Tester which is available at Home Depot and numerous online outlets. Below is a picture of the tester for your reference. This device allows you to determine where each coax cable starts and ends. You place the colored caps on the ends of each coax outlet in your house, then connect the tester to the different connections at the SWM or splitter. This will allow you to use the existing cable to connect to your antenna to all your televisions.

Since you plan to have multiple televisions, you will need to replace the SWM or splitter with a powered amplifier. Channel Master sells 2, 4, or 8 port powered amplifiers on their website, and I highly recommend them. I used a 4 port amplifier outdoors and another 2 port amplifier indoors. So, depending on how many televisions you plan to connect, purchase the correct size amplifier. Below is a picture of a 4 port amplifier. These amplifiers are powered over coax cable as well, so you can use an existing cable to power the amplifier. This enables you to avoid running a power cable to the device if there is no outlet close to the cabling.

The diagram below shows a typical setup with a single antenna, connected to 4 televisions through a powered amplifier. Keep in mind, this can be with an indoor or outdoor antenna. When using an outdoor antenna it is easy to run a new coax cable to the power amplifier from the antenna.

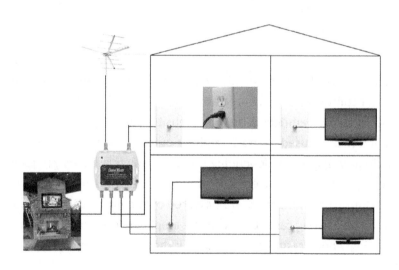

If you chose to install your antenna indoors, such as in the attic and shown in the picture below, you can use the existing cables to connect the antenna to the powered amplifier. You may have to run a new piece of coaxial cable to connect the antenna to the nearest existing coaxial cable inside the house.

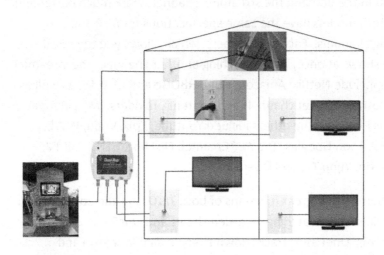

By using one large antenna and a powered amplifier you will be able to have good reception at all your television locations. This also allows you to easily add additional televisions. You may have to add another powered amplifier to ensure any signal lose is overcome. This would be installed where ever you split the current line, or you can upgrade the current powered amplifier to more outputs.

Broadcast Television DVRs

Now that you have broadcast channels at all your television locations, you may want to add a DVR for recoding shows. There are a few options available and each are reviewed below.

Chapter 4: Tivo DVR

Tivo was one of the original DVRs for cable television and they have been a leading inventor of DVR technology ever since. Currently, Tivo has two DVRs available for cord cutters to record OTA broadcast television: the Roamia OTA and the Bolt. I have used both and recommend them as a DVR, but if you plan to use the built in Internet streaming channels, I recommend the Bolt. The hardware is vastly improved in the Bolt and the streaming channels work much better and faster. Both models have the same specifications for DVR and streaming channels. Each has 4 tuners, which allows you to record multiple shows at once, while watching another channel. The streaming channels include Netflix, Amazon, Hulu, HBOGo and MLB.TV, as well as multiple other internet channels and content providers. My personal favorite is Plex. But, the others include Youtube, Epix, Vudu, WWE, Ameba TV, Toon Googles, The ALT Channel, FlixFling, HSN, Tubi TV, Vevo, Yahoo, Yupp TV, and Opera TV.

Some other great functions of both TIVOs are the ability to skip commercials while watching recorded shows and OnePass. According to Tivo.com, "OnePass™ tracks down every available season and episode of a show—whether it's on television or a streaming app—and creates a customizable watchlist for you. Start with the pilot episode or jump in anywhere mid-season. OnePass automatically records upcoming episodes and adds them to your list." This is great for binge watching and catching up on missed episodes. Now you don't have to search across multiple channels and sites to find all the episodes of a show, Tivo does it for you. Another great feature of the Tivo is the ability to watch recorded shows via the Internet. So, you can access your recorded shows, as well as live TV on your phone, tablet or computer. To watch live TV, the Tivo records your show and immediately broadcasts it to you via the Internet. So there is a short delay of a few seconds.

There are two common complaints with Tivo. The first is they charge $15 per month for their service. Which contradicts the reason you are cutting the cord. I chose to pay the fee the first year, and then I upgraded from the Roamio OTA to the Bolt with a lifetime subscription. So, there are ways to avoid the monthly fee. You can pay annual, $149, or purchase a Tivo with a lifetime subscription. The price varies for lifetime subscriptions, but Tivo often has sales on their website. Recently the Bolt was $499 with a lifetime subscription. The Roamio OTA is usually $399 with lifetime subscription. Tivo stopped selling the Roamio OTA with monthly subscriptions. However, these devices can be found on Ebay and Amazon for under $100 with the monthly subscription option.

The second complaint I hear about Tivo is the lack of a Sling TV streaming application. Sling TV is a live TV streaming application sold by Dish Network. It was one of the first live TV streaming application with multiple cable channels, including ESPN, CNN, and AMC. So, many cord cutters chose to use Sling TV for cable channels. With no contract and only $20 per month for the basic package it is very popular. While this is a valid concern, it can be overcome. You can always use a second device such as a Roku to watch Sling, or the new Hulu Live TV is an alternative to Sling and may be available on Tivo soon, given there is already a Hulu app on the Tivo.

Finally, the Tivo interface and remote are the closest option you can get to cable or satellite TV. And with the streaming services built into the Tivo, you don't have to switch from OTA broadcast to a streaming device, such as a Roku or Apple TV. Everything is available in one place. Tivo also connects to the Internet via wired or wireless connections, which makes setup very simple. If you are looking for a smooth, simple operation without having to move between multiple devices, Tivo is a great option.

Chapter 5: Channel Master DVR

Channel Master also makes a DVR for over the air broadcasts. The DVR+ is the second generation of Channel Master's DVR and it is available in 2 models. The only difference between the 2 models is a built in hard drive. So you can pay less and add an external hard drive, or chose the model with a 1TB hard drive included. The cost difference is $249 vs $399 for the internal hard drive model.

The Channel Master DVR+ has two tuners, compared to the Tivo's four, and includes four streaming applications: Sling TV, Youtube, Vudu, and Pandora. Channel Master also has their own streaming channels, called Channel Master TV and currently has over 30+ Internet channels. The only reason I did not chose Channel Master over Tivo, was I wanted a Plex app included on the device. This was more important to me, then a Sling TV application because I am not a heavy cable television viewer. However, I do subscribe to Sling at different times throughout the year and I use a Roku device to watch Sling TV. If you are a heavy cable watcher, then the available Sling TV application on the Channel Master DVR+ may be the right choice for you. Also, there is no monthly fee with the Channel Master DVR+, which is a plus compared to the Tivo.

One common complaint with the DVR+ is there is no built in WiFi, so you have to purchase a $39 USB WiFi adapter from Channel Master. While this is not a big deal, I hope Channel Master's next generation DVR will have built in WiFi.

Based on internet reviews and feedback, the majority of users are happy with the Channel Master DVR+. It is a difficult choice between the Tivo and Channel Master, but my recommendation is purchase a Tivo Bolt with lifetime subscription.

Chapter 6: Tablo DVR

The third option for OTA DVRs is a little different then the Tivo and Channel Master DVRs. The Tablo DVR is a whole house DVR that connects directly to your antenna and rebroadcasts the live TV and recorded shows to your streaming devices, computer, tablets, and cell phones. So, rather than having a TIvo or Channel Master DVR+ at your television, you need a Roku, Apple TV, Chromecast, or some other streaming device to watch the Tablo re-broadcasted live TV on your television. You control the Tablo DVR through the streaming device and can watch live, or recorded shows.

There are a few distinct advantages of the Tablo over the Tivo and Channel Master. First, you do not need to have coxial cable run to your televisions. Rather, the signal is broadcast via WiFi from the Tablo to your streaming device that is connected to your television. This can be significant if you do not already have coaxial cable to your television location. Second, you can watch live television and all your recordings on your computer, phone or tablet. Finally, you only need one streaming device, such as a Roku connected to your television. You can watch live television through the Tablo app, and also access all your other streaming services, such as Netflix on the same device. So, no switching between inputs or using multiple remote controls.

For installation of the Tablo DVR, you only need to connect the Tablo to your antenna and a hard drive to store your recordings. Everything else is connected via WiFi, or Ethernet. The WiFi option for the Tablo makes installation very simple. You only have to worry about connecting the Tablo to the antenna, and then everything can be operated via wireless WiFi connections. Tablo has both 2 and 4 turner models available, and recently released 2 new devices. The first is a new 2 tuner DVR with 64GB of built in storage. This allows you to get up and running without an external hard drive.

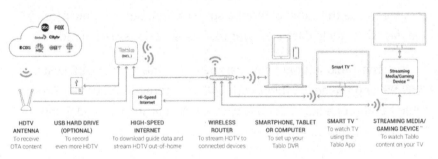

The second is a USB tuner and application for the Nvidia Shield streaming device. This allows you to watch and record shows directly to the Nvidia Shield connected to your television. The downside to the

Nvidia Shield setup, is you need a wired connection between the antenna, USB tuner and Nvidia Shield, which is connected to your television. But, for a small apartment or single TV setup, this will work fine.

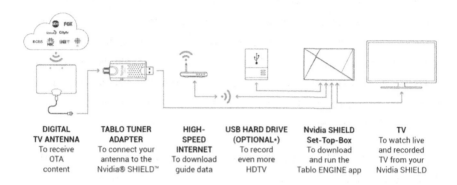

DIGITAL TV ANTENNA	TABLO TUNER ADAPTER	HIGH-SPEED INTERNET	USB HARD DRIVE (OPTIONAL*)	Nvidia SHIELD Set-Top-Box	TV
To receive OTA content	To connect your antenna to the Nvidia® SHIELD™	To download guide data	To record even more HDTV	To download and run the Tablo ENGINE app	To watch live and recorded TV from your Nvidia SHIELD

There are many happy customers of the Tablo and their development has continued since launch in 2014. I like the idea of being able to watch live television on multiple devices via WiFi and the setup is very simple. One item to note, similar to TIvo, Tablo does charge a monthly, annual, or lifetime subscription fee for their guide listings. The costs are $4.99, $49.99, or $149.99 respectively. It is possible to use the Tablo without a subscription, but you only receive one day of TV guide data and your recording options are limited.

The Tablo 2 tuner DVR costs $219.99, the 4 tuner DVR costs $299.99, and the new 2 tuner with 64GB of storage is $219.99. The Tablo USB tuner adapter for use with the Nvidia Shild is $69.99 and the Tablo engine application is free.

Chapter 7: HDHomeRun

The HDHomeRun Connect by Silicon Dust is another whole home DVR that operates like the Tablo DVR. The HDHomeRun Connect is connected to the antenna and then the live television is rebroadcast via WiFi or Ethernet cable to your television and devices. One difference with the HDHomeRun Connect, the hard drive for recoding the television is not connected to the device, but is a network attached storage (NAS) device or local computer. So while the HDHomeRun connect is only $99, you will need to purchase a NAS storage device or configure your local computer to function as a DVR. Also, the HDHomeRun Connect must be connected via Ethernet to the router, while the Tablo can connect via Ethernet or WiFi to the router. For anyone who uses Plex, the HDHomeRun Connect is compatible with Plex and allows you to use Plex as your DVR. So, your home computer running Plex will serve as your DVR. There is also a $35 per year fee for the DVR subscription and 14 day TV Guide listings. But, as with the Tablo, you can use the HDHomeRun Connect without a subscription. However, you only receive 24 hours of TV Guide listings and your recording options are reduced.

If you are a Plex user, the HDHomerun Connect maybe a good option. Otherwise, the Tablo may be easier to setup and configure.

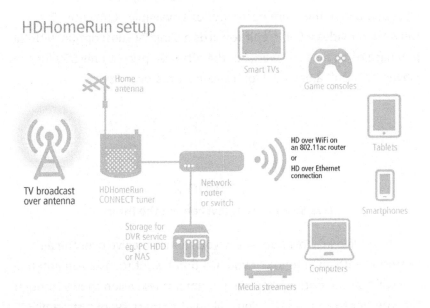

Chapter 8: AirTV Player

Dish Network introduced the AirTV Player and AirTV Adapter as a stand alone box to watch Sling TV, while allowing access to other streaming apps as well. The AirTV Player runs the Android TV operating systems, but instead of booting into the Android menu, AirTV launches into a Sling TV app. If you add the AirTV adapter to the Player, you are able to connect an over the air antenna, and the available local live channels are integrated in to the Sling TV app. This is a great concept, but unfortunately the AirTVPlayer and adapter need improvements. The guide section of the Sling TV app is very cumbersome and is not easy to use. There is no traditional grid based channel guide, so determining what is playing on each channel at different times is difficult. In addition to Netflix preinstalled, you are able to add streaming apps from the Google store since the Air TV player runs Android.

Unfortunately, the AirTV Player lacks a DVR, so you cannot record over the air broadcasts. This is a serious limitation compared to other OTA options. While the price is only $99 (player), $39 (adapter), or $129 combined, there are better options available. Currently, Dish Network provides a $50 credit towards a Sling TV subscription with the purchase of an Air TV Player. So, the effective price is only $50 for the player. This isn't a bad deal for a second streaming device.

Section 2
Live Streaming Television via the Internet

If you live in a location where you cannot receive over the air television with an antenna or you just don't want to install an antenna, you still have an option. You can stream live television shows through the Internet to your home. You will need a Smart TV, or a streaming device to watch the shows on your television. You can also watch many of the shows on your computer, tablet, or smart phone. This section will describe each live TV streaming service and at the end is a table that compares what channels are available on each service. When making a choice of live streaming television providers, it is important to review the available channels to ensure you will have access to the channels and shows you watch. It is also important to ensure you have the streaming device that works with the provider. Every streaming provider is not available on every device, so review the table of devices and providers as well. Another important factor when choosing which streaming provider to pick, is the available on demand options and cloud DVR service. If you currently record many of your favorite shows and do not watch them live, then you need a service with DVR or on demand access to your shows and channels. Otherwise, you have to watch the shows live, or not at all.

Chapter 9: Sling Television

Sling was the first live television streaming provider and is also the most popular provider. Sling is owned by Dish Network and was first launched in February 2016. In June 2017, Sling reached 2 million subscribers. There are no contracts and the monthly cost starts at $20 for the orange package of 28 channels, $25 for 44 channels in the blue package, or $40 for all the channels. There are also 30 additional add-on packages such as sports extra, kids extra, Hollywood extra, world news extra, and best of Spanish TV. The add-on packages also include HBO, Showtime, Cinemax, and Starz. Each add-on package costs between $5 and $15. Including the channels available with the add-on packages, Sling offers more than 150 channels. As you can see, you can quickly approach the cost of normal cable/satellite television. So, you must decide what channels are most important and pick the package that works best for you. Below is a comparison of the Sling packages. All channels are included with the blue and orange package. The main difference is Sling Orange includes ESPN, ESPN2, ESPN3, and the Disney Channel. While Sling Blue includes Bravo, Fox Regional Sports, FX, and the NFL Network.

Orange vs. Blue Sling Packages

CHANNEL	ORANGE	BLUE
A&E	✓	✓
AMC	✓	✓
AXS TV	✓	✓
BBC America	✓	✓
Bloomberg TV	✓	✓
Cartoon Network / Adult Swim	✓	✓
Cheddar	✓	✓
CNN	✓	✓
Comedy Central	✓	✓
Flama	✓	✓
Food Network	✓	✓
Galavision	✓	✓
HGTV	✓	✓
History	✓	✓

IFC	✓	✓
Lifetime	✓	✓
Local Now	✓	✓
Newsy	✓	✓
TBS	✓	✓
TNT	✓	✓
Travel Channel	✓	✓
Viceland	✓	✓
Disney Channel	✓	
ESPN	✓	
ESPN2	✓	
Freeform	✓	
Fox		✓
NBC		✓
Bravo		✓
Fox Sports regionals		✓
FS1		✓

FS2	✓
FX	✓
FXX	✓
National Geographic	✓
Nat Geo Wild	✓
NBCSN	✓
NFL Network	✓
Syfy	✓
USA Network	✓
BET	✓
Unimas	✓
Univision	✓
TruTV	✓

Sling does provide a DVR service for $5 per month and includes 50 hours of recording time. In addition to the DVR, Sling also provides on-demand access to content on many of their channels. The content and length of time they are available varies by channel, but I have found most popular cable shows have at least the last few shows available. But, this is not a binge option like Netflix, HBO, etc.

Recently Sling has begun to offer local channels: FOX, NBC, ABC, Univision, and Unimas in over 50 markets. To check what is available in your area goto, http://help.sling.com/articles/en_US/FAQ/How-do-local-channels-work-on-Sling-TV and enter your zip code. Sling uses your billing zip code to determine which local channels you will receive. But, the current list as of this books release is below.

Market	ABC	FOX	NBC	Univision	Unimas
Chicago	✓	✓	✓	✓	✓
New York	✓	✓	✓	✓	✓
San Francisco	✓	✓	✓	✓	✓
Los Angeles	✓	✓	✓	✓	✓
Philadelphia	✓	✓	✓	✓	National
Houston	✓	✓	On Demand	✓	✓
Dallas		✓	✓	✓	✓
Washington D.C.		✓	✓	✓	National
Fresno/Visalia	✓	On Demand	On Demand	✓	✓
Raleigh	✓	On Demand	On Demand	✓	National
Orlando		✓	On Demand	✓	✓
Tampa		✓	On	✓	✓

			Demand		
Phoenix		✓	On Demand	✓	✓
Atlanta		✓	On Demand	✓	National
Austin, TX		✓	On Demand	✓	National
Gainesville		✓	On Demand	National	National
Minneapolis		✓	On Demand	National	National
Detroit		✓	On Demand	National	National
Charlotte		✓	On Demand	National	National
Miami		On Demand	✓	✓	✓
San Diego		On Demand	✓	✓	✓
Hartford		On Demand	✓	✓	National
Boston		On Demand	On Demand	✓	✓
Harlingen		On Demand	On Demand	✓	✓

San Antonio	On Demand	On Demand	✓	✓
Denver	On Demand	On Demand	✓	✓
El Paso	On Demand	On Demand	✓	✓
Yuma/El Centro	On Demand	On Demand	✓	✓
Albuquerque	On Demand	On Demand	✓	✓
Sacramento	On Demand	On Demand	✓	✓
Cleveland	On Demand	On Demand	✓	National
Fort Myers	On Demand	On Demand	✓	National
Corpus Christi	On Demand	On Demand	✓	National
Odessa/Midland	On Demand	On Demand	✓	National
Oklahoma City	On Demand	On Demand	✓	National
Wichita	On Demand	On Demand	✓	National
Laredo	On Demand	On Demand	✓	National

Colorado Springs	On Demand	On Demand	✓	National
Salt Lake City	On Demand	On Demand	✓	National
Tucson	On Demand	On Demand	✓	National
Bakersfield	On Demand	On Demand	✓	National
Palm Springs	On Demand	On Demand	✓	National
Yakima/Pasco	On Demand	On Demand	✓	National
Reno	On Demand	On Demand	✓	National
Seattle/Tacoma	On Demand	On Demand	✓	National
Portland	On Demand	On Demand	✓	National
Monterey	On Demand	On Demand	✓	National
Las Vegas	On Demand	On Demand	✓	National
Santa Barbara	On Demand	On Demand	✓	National

Sling is available on almost every device imaginable. From your computer to your phone, or Roku player. Below is a list of all devices that support Sling TV. The only non supported device compared to other services is the Playstation 4, which only supports Playstation Vue.

Sling Supported Devices

Apple TV	Amazon Fire TV
Roku	Chromecast
Android TV	Xiaomi
Leeco	AirTV Player
ZTE	Channel Master
LG WebOS	IOS iPhone
Android Phones	Amazon Fire Phones
Xbox One	OSX (Mac/apple computers)
Windows 10 (PC Computers)	

Chapter 10: PlayStation Vue

On March 18, 2015, Sony released PlayStation Vue as a direct competitor to Sling TV. PS Vue has 400,000 subscribers as of March, 2017 and offers packages that start at $39.99 and go up with premium add ons. One advantage of PlayStation Vue is it provides on-demand access to ABC, CBS, NBC, and FOX shows. So, even though you may not be able to watch the shows live, you will have access to new shows after 24 hours. Not a bad compromise compared to cable. The 4 different packages and add ons are provided in the tables below. The cheapest package, Access has 53 channels, which includes the on-demand channels of ABC, CBS, NBC, and FOX. So, there are actual 49 live channels. Which is similar in cost and available channels to Sling TV. 49

(PS Vue) vs 48 (Sling) both for $40. So, in my opinion the access package on PS Vue is a better deal than the blue/orange combined package on Sling TV. Simply because you receive the on demand access to ABC, CBS, NBC, and FOX.

PS Vue also provides free DVR as part of every package. However, the recordings are only available for 28 days. The recordings can be watched via mobile devices and Internet Web browers from anywhere, which is a nice feature.

PS Vue also provides some live local television in select markets. Over 90% of CBS affiliates are available on PS Vue, and recent agreements by ABC, NBC, and FOX with local affiliates will allow PS Vue and other streaming services to begin to offer more local live television. In the meantime, PS Vue does offer ABC, NBC, and FOX on-demand shows. So, while you may not be able to watch local live news or sporting events, you can watch your favorite sitcoms from ABC, NBC, and FOX.

PS Vue is available on multiple devices including, Apple TV, Amazon Fire TV, Roku, Chromecast, Android TV, IOS iPhones, Android Phones, OSX (Mac/apple computers), Windows 10 (PC Computers), PlayStation4.

PlayStation Vue Packages

Channel	Ultra $74.99	Elite $54.99	Core $44.99	Access $39.99
ABC On Demand	X	X	X	X
CBS (Some Markets)	X	X	X	X
FOX / FOX On Demand	X	X	X	X
NBC On Demand	X	X	X	X
AdultSwim	X	X	X	X
AMC	X	X	X	X
AHC	X	X		
Animal Planet	X	X	X	X

BBC America	X	X	X	X
Boomerang	X	X		
Bravo	X	X	X	X
BTN	X	X	X	
Cartoon Network	X	X	X	X
Chiller	X	X		
CNBC	X	X	X	X
CNBC World	X	X		
CNN	X	X	X	X
Comcast Sports Net (regional)	X	X	X	
Comcast Network (regional)	X	X	X	
Cooking Channel	X	X		
Destination America	X	X	X	X
Discovery	X	X	X	X
Discovery Family	X	X	X	X
Discovery Life	X	X		
Disney	X	X	X	X
Disney Jr.	X	X	X	X
Disney XD	X	X	X	X
DIY	X	X	X	X
E!	X	X	X	X
EPIX Hits	X	X		
ESPN	X	X	X	X
ESPN Deportes	X	X	X	
ESPN2	X	X	X	X
ESPNEWS	X	X	X	
ESPNU	X	X	X	
Food Network	X	X	X	X
Fox Business	X	X	X	X
Fox College Sports Atlantic	X	X		
Fox College Sports Central	X	X		
Fox College Sports Pacific	X	X		

Fox Deportes	X	X		
Fox News	X	X	X	X
Fox Sports Net (regional)	X	X	X	
Fox Sports 1	X	X	X	X
Fox Sports 2	X	X	X	X
Freeform	X	X	X	X
Fusion	X	X		
FX	X	X	X	X
FXM	X	X		
FXX	X	X	X	X
Golf	X	X	X	
HBO	X			
HGTV	X	X	X	X
Hi-YAH	X	X		
HLN	X	X	X	X
IFC	X	X	X	
Impact	X	X		
Investigation Discovery	X	X	X	X
Longhorn Network (regional)	X	X	X	
Machinima	X	X		
MLB TV	X	X	X	
MLB TV Alternate	X	X	X	
MGM	X	X		
MSNBC	X	X	X	X
Nat Geo Wild	X	X		
Nat Geo	X	X	X	X
NBA TV	X	X	X	
NBC Sports	X	X	X	X
NFL Network	X	X	X	
One World Sports	X	X		
Outside	X	X		
OWN	X	X	X	X

Oxygen	X	X	X	X
Poker Central	X	X		
Pop	X	X	X	X
Science Channel	X	X	X	X
SEC Network	X	X	X	
Showtime	X			
Sony Movie Channel	X	X		
Spike	X	X	X	X
Sprout	X	X		
Sundance	X	X	X	
Syfy	X	X	X	X
TBS	X	X	X	X
Telemundo	X	X	X	X
TLC	X	X	X	X
TNT	X	X	X	X
Travel Channel	X	X	X	X
TruTV	X	X	X	X
TCM	X	X	X	
Universal	X	X		
USA	X	X	X	X
Velocity	X	X		
WeTV	X	X	X	X

PlayStation Vue offers the following addon packages as well.

Fox Soccer Plus	$15/month
NFL RedZone	$40/season
HBO (live linear feeds and On Demand)	$15/month
Cinemax (live linear feeds and On Demand)	$15/month

Showtime (live linear feeds and On Demand)	$11/month
Epix Hits (live linear feed and On Demand)	$4/month
Español Pack (Cine Sony Television, CNN en Español, Discovery en Español, Discovery Familia, Fox Deportes, Fox Life, Nat Geo Mundo, NBC Universo)	$5/month
Machinima	$2/month
Polaris TV	$3/month

Chapter 11: Hulu Live TV

In May 2017, Hulu launched a new live television service with 54 channels for $39.99 per month. There are 3 add-on channel options (HBO, Cinemax, Showtime) that cost $14.99, $9.99, and $8.99 respectively. In addition to the live broadcasts, you also receive the standard Hulu streaming library of on-demand shows at no additional costs. Hulu standard costs $7.99 per month as a standalone service. You receive 50 hours of cloud DVR storage, which can be upgraded to 200 hours of storage for $14.99 per month. The base service allows 2 devices to stream shows simultaneously, and for an additional $14.99 per month, you can stream to unlimited devices.

The 54 channels include the national broadcasts of ABC, NBC, CBS, and FOX, as well as most major cable networks. A few notable exceptions are AMC, Comedy Central, Discovery Channel, MTV, Spike and Nickelodeon.

Hulu Live TV does offer some support for live local stations, but it is hit and miss. However, as with the other services, more are being added all the time. All four (ABC, CBS, NBC, FOX) local affiliates are available in New York, San Francisco, Philadelphia, Los Angeles, and

Chicago.

Sports channels include the ESPN channels, SEC channel, Big Ten Network, Golf Channel, and NBC Sports Network. Some markets are also available for regional sports on NBC Sports Network and Fox Regional Sports Networks.

Hulu Live TV Channels

ABC	CBS	Fox
NBC	A&E	Big Ten Network
Boomerang	Bravo	Cartoon Network
CBS Sports Network	Chiller	CNBC
CNN	CNN International	Disney Channel
Disney Junior	Disney XD	E!
ESPN	ESPN 2	ESPN News
ESPN U	Food Network	Fox Business
Fox News	FS1	FS2
Free Form	FX	FXM
FXX	FYI	NBC Golf
HGTV	History Channel	HLN
Lifetime	LMN	MSNBC
National Geographic	Nat Geo Wild	NBCSN
Oxygen	Pop	SEC Network
Sprout	SyFy Channel	TBS

Turner Classic Movies	TNT	Travel Channel
Tru TV	USA	Viceland

FOX Regional Sports Networks   **NBC Sports Networks**

FOX Regional Sports Networks	NBC Sports Networks
FS Arizona	NBCSN Bay Area
FS Detroit	NBCSN California
FS Florida	NBCSN Chicago
FS Midwest	NBCSN Mid-Atlantic
FS North	NBCSN New England
FS Ohio	NBCSN Northwest
FS Prime Ticket	NBCSN Philadelphia
FS San Diego	
FS South	
FS Southeast	
FS Southwest	
FS Sun	
FS West	
SportsTime Ohio	
YES	

A few draw backs to Hulu Live TV are the inability to fast forward through commercials on both DVR recordings and on-demand shows. If you pay for the enhanced DVR feature you are able to skip commercials. There is also a $4 per month upgrade to remove ads and commercials from the on-demand shows. Also, Hulu Live TV is not available on Roku devices nor Mac and PCs. But, it does support Apple TV, Amazon Fire, Chromecast, Xbox, as well as Apple and Android Phones and tablets. Hulu says support for Roku, Samsung, Mac and PCs is coming soon.

The biggest advantage of Hulu Live TV, is the blended live TV and on demand content into a single service. They currently offer a 7 day trial, and it is worth trying out to see if you like it.

Chapter 12: YouTube TV

Not be left out of anything technology related, Google has joined the streaming of live television with YouTube TV. For $35 a month, you get 48 channels of live TV and unlimited DVR storage. YouTube TV membership comes with 6 accounts and allows 3 simultaneous streams.

While the price of YouTube TV, unlimited DVR, and 6 streaming accounts are great, YouTube TV is lacking in a few areas. The first is the channels available. There are no Viacom channels (Comedy Central, MTV, Spike, Nickelodeon), no Turner channels (TNT, TBS, CNN), no Discovery channels, no A&E, and no HBO. Second is the lack of device support for YouTube TV. Chromecast and Apple Air Play are the only streaming devices that support YouTube TV for streaming content to your television. But you can watch YouTube TV on both Android and Apple phones, as well as your computer via the browser.

The biggest problem with YouTube TV at the time of this writing is, it is only available if you live in 29 metro areas. So if you don't live in one of these areas, you cannot sign up for YouTube TV.

YouTube Live TV Localities

Atlanta	Baltimore
Boston	Charlotte
Chicago	Cincinnati
Columbus	Dallas-Fort Worth
Detroit	Houston
Jacksonville	Las Vegas
Los Angeles	Louisville
Memphis	Miami-Fort Lauderdale
Minneapolis-Saint Paul	Nashville
New York City	Orlando-Daytona Beach-Melbourne
Philadelphia	Phoenix
Pittsburgh	San Antonio
San Francisco Bay Area	Seattle
Tampa	Washington DC
West Palm Beach	

The sports and news offerings on YouTube TV are excellent and you receive a free Chromecast with your first payment. Google also offers a one-month free trial. If you can live without the missing cable channels, don't mind using a Chromecast to watch TV, and you live in one of the 29 metro areas, then YouTube TV maybe a good option. If history is any judge, Google will steadily increase support to YouTube TV and role the service out nationwide. I look forward for more to come from YouTube TV.

YouTube TV Channels

Broadcast	Entertainment	Sports	News	Kids	Spanish
ABC	AMC	Big Ten	BBC News	Disney Channel	Telemundo
CBS	BBC America	CBS Sports	CNBC	Disney Jr	Universo
The CW	Bravo	Comcast RSN	Fox Business	Disney XD	
FOX	Chiller	ESPN	Fox News	Sprout	
NBC	E!	ESPN2	MSNBC		
	Esquire	ESPNews			
	Freeform	ESPNU			
	FX	Fox RSN			
	FXM	FS1			
	FXX	FS2			
	IFC	Golf Channel			
	Nat Geo	NBCSN			
	Nat Geo Wild	SEC Network			
	Oxygen				

	Sundance				
	SyFy				
	Universal HD				
	USAWE tv				
	YouTube Red Originals				

Chapter 13: Direct TV Now

Satellite TV operator Direct TV (owned by AT&T), launched a streaming TV service, Direct TV Now in November 2016. The service has 4 different packages, starting at $35 for 60 channels, and increasing up to $70 for 129 channels. The middle packages are $50 and $60 for 80 and 100 channels, respectively. Each account is allowed 2 simultaneous streams, and the service can be watched outside the home.

The channel selection is excellent and is very similar to standard cable or satellite packages, and Direct TV Now is the cheapest option on a per channel option. Notable channels not available are Showtime and the NFL Network. HBO, Cinemax, and Starz are available as add-on packages.

There is currently no DVR capability with Direct TV Now. However, a DVR beta program was recently launched, so the service may be coming soon to all users. Direct TV Now does offer 72 hour replay, so the majority of shows are available for replay 72 hours after airing. While you can't manage recordings and keep them longer than 72 hours, you do have the ability to watch shows you missed.

Live local television is available in some markets, but there is no live CBS broadcasts. You can check if local channels are available in your zip code here, http://www.directv.com/DTVAPP/packProg/attLocalChannels.jsp. The big four broadcast channels, ABC, CBS, NBC, and FOX, are available on-demand, so you can watch your favorite shows the day after they air.

One major advantage of Direct TV Now is for customers of AT&T mobile or home Internet service. If you have an AT&T account, you can add Direct TV Now for $10, after a $25 rebate. You can also stream television to your mobile devices and it doesn't count against your data cap.

Direct TV Now is available on the following devices. The only major device missing compared to other services is Android TV.

- Amazon Fire TV
- Apple TV
- Roku
- Chromecast
- Google-cast enabled LeEco TVs
- Vizio SmarrtCast displays
- iOS devices (iPhone/iPad)
- Android devices (phones/tablets)
- Web browsers

Direct TV Now is the closet streaming service to standard cable or satellite television. But, there are no contracts and you can watch on your mobile devices. The $35 package is a good deal, but as the price increases you are close to cost of normal cable. Hopefully, DVR service will be coming soon to everyone and local live television broadcasts will continue to increase. If you are an AT&T customer, the $25 rebate per month is a great deal, and it is worth trying. Otherwise, try the 7 day free trial and see what you think.

Direct TV Now Channel Listings by Package

	Live a Little	Just Right	Go Big	Gotta Have It
	$35	$50	$60	$70
	60+ channels	80+ channels	100+ channels	120+ channels
A Wealth of Entertainment		X	X	X
A&E	X	X	X	X
ABC (on-demand)	X	X	X	X
AMC	X	X	X	X
American Heroes			X	X
Animal Planet	X	X	X	X
Audience	X	X	X	X
AXS TV	X	X	X	X
Baby First	X	X	X	X
BBC America	X	X	X	X
BBC World News			X	X
BET	X	X	X	X
Bloomberg TV	X	X	X	X
Boomerang				X
Bravo	X	X	X	X
BTN		X	X	X
C-SPAN	X	X	X	X

C-SPAN2	X	X	X	X
Cartoon Network	X	X	X	X
Centric			X	X
Chiller				X
CMT	X	X	X	X
CNBC	X	X	X	X
CNBC World		X	X	X
CNN	X	X	X	X
Comedy Central	X	X	X	X
Comedy TV		X	X	X
Cooking Channel		X	X	X
Destination America			X	X
Discovery	X	X	X	X
Discovery Family Channel			X	X
Discovery Life			X	X
Disney Channel	X	X	X	X
Disney Junior	X	X	X	X
Disney XD	X	X	X	X
DIY Network			X	X
E!	X	X	X	X
El Rey				X

ESPN	X	X	X	X
ESPN2	X	X	X	X
ESPNEWS		X	X	X
ESPNU		X	X	X
FM			X	X
Food Network	X	X	X	X
Fox (on-demand)	X	X	X	X
Fox Business Network	X	X	X	X
Fox News Channel	X	X	X	X
FOX Sports 1	X	X	X	X
FOX Sports 2			X	X
Freeform	X	X	X	X
Fuse		X	X	X
Fusion		X	X	X
FX	X	X	X	X
FX Movie Channel			X	X
FXX	X	X	X	X
fyi			X	X
Galavision	X	X	X	X
Golf Channel			X	X
GSN		X	X	X

Hallmark Channel	X	X	X	X
Hallmark Movies & Mysteries	X			X
HGTV	X	X	X	X
HISTORY	X	X	X	X
HLN	X	X	X	X
IFC		X	X	X
Investigation Discovery	X	X	X	X
Justice Central		X	X	X
Lifetime	X	X	X	X
LMN			X	X
Logo			X	X
MLB Network		X	X	X
MSNBC	X	X	X	X
MTV	X	X	X	X
MTV2	X	X	X	X
MTV Classic			X	X
Nat Geo WILD			X	X
National Geographic Channel	X	X	X	X
NBA TV			X	X
NBC (on-demand)	X	X	X	X

NBCSN	X	X	X	X
NHL Network			X	X
Nick Jr.	X	X	X	X
Nickelodeon/Nick at Nite	X	X	X	X
Nicktoons		X	X	X
One American News	X	X	X	X
OWN		X	X	X
Oxygen			X	X
Revolt			X	X
RFD-TV	X	X	X	X
Science		X	X	X
SEC Network		X	X	X
Spike	X	X	X	X
Sprout			X	X
STARZ ENCORE Action				X
STARZ ENCORE Black				X
STARZ ENCORE East				X
STARZ ENCORE Family				X
STARZ ENCORE Classic				X
STARZ ENCORE Suspense				X

STARZ ENCORE West				X
STARZ ENCORE Westerns				X
SundanceTV		X	X	X
SYFY	X	X	X	X
TBS	X	X	X	X
TCM	X	X	X	X
TeenNick	X	X	X	X
Telemundo (on-demand)	X	X	X	X
Tennis Channel		X	X	X
TLC	X	X	X	X
TNT	X	X	X	X
Travel Channel		X	X	X
TruTV	X	X	X	X
TV Land	X	X	X	X
TVG			X	X
TV One		X	X	X
UniMás		X	X	X
Universo			X	X
Univision	X	X	X	X
Univision Deportes Network				X

USA Network	X	X	X	X
Velocity	X	X	X	X
VH1	X	X	X	X
Viceland	X	X	X	X
WE tv	X	X	X	X
Weather Channel		X	X	X
WeatherNation	X	X	X	X

Chapter 14: Fubo TV

Another live television service that is often over looked is Fubo TV. Fubo TV is more focused on sports and international programming than the other services. Fubo TV has 3 basic packages and 7 add-ons available. The standard package is called Fubo Premier and it includes 63 channels for $34.99 per month. Of those 63 channels, 23 are sports channels and 18 are international channels. The other 2 packages foreign language packages, Fubo Latino and Fubo Portugues.

Fubo Latino provides 13 Spanish language channels for $14.99 per month. While, Fubo Portugues provides 5 Portuguese language channels for $19.99 per month. Both Fubo Latino and Portugues packages are available as add-on packages as well.

Additional add-on packages are Sports Plus, Mundo Plus, The Blues, Cycling Plus and Kids Plus. There is also the Entretenimento Plus add-on package that is only available with the Fubo Latino base package.

Fubo TV offers a cloud based DVR service and a 72 hour loop back service that allows on-demand access to most content for 72 hours.

Fubo TV is available on multiple devices including, Apple TV, Amazon Fire TV, Roku, Chromecast, Android TV, IOS iPhones, Android

Phones, OSX (Mac/apple computers), and Windows 10 (PC Computers).

Although Fubo is a sports focused service, there is no ESPN or NFL Network. If you are looking for more sports options and Spanish language television, then Fubo TV maybe a good option. They offer a 7 day trial period.

Fubo Premier Channels

A&E	Bein Sports (English)
Big Ten Network	Bein Sports (Spanish)
Chiller	Bravo
Eleven Sports	CNBC
Football Report TV	E!
Fox Business Channel	El Rey
Fox Regional Sports Networks	Fox (some locations)
Comcast Sports (some locations)	Fox News Channel
FS1	Fox Soccer Plus
Fuse	FS2
FXM	FX
FYI	FXX
Golf Channel	Hallmark Channel
Hallmark Movies and Mysteries	History
Lifetime	Lifetime Movies
Local Now	MSNBC

MTV	Nat Geo
Nat Geo Wild	NBC (some locations)
NBA TV	NBC Sports Network
Oxygen	Revolt
SyFy	Sprout
Telemundo (some locations)	Universal
Telexitos (some locations)	Univision Deportes
Unimas	Universo (NBC)
Univision	USA
Viceland	

Fubo Latino Channels

Bein Sports English	Baby TV Spanish
Cine Sony Television	Bein Sports Spanish
Fox Life	Fox Deportes
GolTV Spanish	Galavision
TyC Sports	Nat Geo Mundo
Univision	UniMas
Univision Deportes	

Fubo Portugues Channels

Bein Sports English	Bein Sports Spanish
Benfica TV	GolTV Spanish
RTPi	

Fubo Add-ons

- Portuguese Plus: RTPi, Benfica TV, GolTV Spanish

- Sports Plus: GolTV Spanish, GolTV English, TyC Sports, FJ Fight Network, FNTSY Sports Network

- Mundo Plus: Net Geo Nuno, Fox Life, Fox Deportes, TyC Sports, GolTV Spanish, Cine Sony TV

- Fubo Latino: GolfTV Spanish, Fox Deportes, TyC Sports, Fox Life, Nat Geo Mundo, Cine Sony TV, Baby TV Spanish

- The Blues: Chelsea TV

- Cycling Plus: Fubo Cycling

- Kids Plus: Baby TV

Chapter 15: Pluto TV

The final live television service is also the only free live television service. Pluto TV offers over 100 free television channels; some you know and many you never heard of. Pluto TV operates like a traditional cable service, in that it is supported by commercial advertising. That is how they offer the service for free to customers.

The channels you may be familiar with include NBC News, MSNBC, CNBC, CBS News, Sky News, The Weather Network, NASA TV, and Bloomberg TV. Other channels are created based on content and many are very good. They include the Surf Network, Poker Tour, Pro

Wrestling, Kickboxing, Food TV, Science TV, Stand Up Comedy, and Fail TV. There are also 3 kids channels, Kids TV, After School Cartoons, and Classic Toons TV. Pluto also provides 32 music channels, which is a nice addition.

Pluto is available on iPhones and Android, Roku, Amazon Fire TV, Apple TV, Android TV, Chromecast, Samsung Smart TVs, Vizio TVs, Sony Smart TVs, and via MAC and PC Computers.

While Pluto TV doesn't have many of the traditional cable channels, nor live local television and ABC, CBS, NBC, and FOX, it is a great free service. If you are looking to have a no cost option, Pluto TV, plus an antenna is the way to go.

Pluto TV Channels

News channels	Sports channels	Entertainment
Todays Top Stories	Fear Factor	Crime Network
News 24/7	Glory Kickboxing	Shout! Factory TV
NBC News/MSNBC	Extreme Sports	Classic TV
Sky News	Surf Channel	Live Music Replay
Business News 24/7	Skate	Kids TV
RT America	Big Sky Conference	After School Cartoons
NewsmaxTV	World Poker Tour	Classic Toons
TYT Network	Fight	Awesomeness TV
The Weather Network	Sports News	Rocket Jump
CNBC	Pro Wrestling	What?!
Cheddar		

CBSN	Comedy	Geek and Gaming channels
Newsy	Funny AF	The Feed
Bloomberg TV	Stand Up	Hive
	Cracked	Anime All Day
Movies	The Onion	Minecraft TV
Pluto TV Movies	MST3K	IGN
Action Movies	Rifftrax	Geek & Sundry
Flicks of Fury	Fail TV	cNet
Horror 24/7		Nerdist
Classic Movies	**Chill Out**	
Gravitas Movies	Internet Gold	**Life + Style**
The Asylum	Vibes	People Entertainment
	Montercat TV	Clevver
Curiosity	THC	Popsugar
Xive TV	Man UP	Front Door
Science TV	Eye Candy	Adventure TV
DocuTV	Revry	Food TV
Nasa TV	Nature Vision TV	Tastmade
	Slow TV	
	4K TV	

Pay Streaming Services (Recorded Movies and Television)

In addition to streaming live television via Internet there are many options for recorded television and movies. These are great options to supplement your live television via antenna or the Internet. In addition to pay services such as Netflix, Hulu, and Amazon Prime, there are free services such as YouTube, Crackle, and Tubi TV. Either alone or bundled with over the air television, there are great options for cord cutters.

Chapter 16: Netflix

The most popular streaming service on the Internet is hands down Netflix. In July 2017, it was announced that Netflix has over 100 million subscribers worldwide, and 50 million in the USA. In addition to thousands of movies and television shows, Netflix also produces original content such as House of Cards, Orange is the New Black, and Stranger Things. Each month new movies and TV shows are added to Netflix, while others are removed. Interestingly, although Netflix is the most popular it does not have the most content. Both Amazon Prime and Hulu have more movies or TV shows. In March 2017, Barclays reported that Amazon Prime Video offered 18,405 movies and 1,981 TV shows in the U.S., while Netflix offered 4,563 movies and 2,445 series. Hulu also tops Netflix in both areas with 3,588 shows and 6,656 movies. Something to consider when deciding on a movie streaming service.

Netflix offers 3 different plans and the table below breaks down the difference by costs, available streams and resolution.

Plan	Price	Number of screens	Resolution
Basic (streaming)	$8	1	SD
Standard (streaming)	$10	2	HD
Premium (streaming)	$12	4	HD + Ultra HD

Netflix is available on just about every platform and device available. The table at the end of this section shows a comparison of platforms and devices versus streaming services.

Chapter 17: Hulu

In 2008, Hulu started as a service to stream television shows the day after they aired on live TV, and also included a catalog of past shows. Originally Hulu offered a free and paid service, but has since changed to only a pay service. Major networks available on Hulu include ABC, NBC, and FOX, as well as many other cable networks. As previously stated, Hulu also has a large catalog of movies.

Hulu has recently begun to produce original content, and has started a live television service. Hulu costs $7.99 per month, and has an add-on service to eliminate all commercials for another $4 per month. Hulu currently offers a 30 day trial. So, if you don't mind watching your TV shows one day late, then this might be a good all around option for TV and movies.

Chapter 18: Amazon Prime

When most people hear Amazon Prime, they immediately think of free 2 day shipping. But, also included with your Amazon Prime membership is a huge catalog of movies and television, as well as original content. Amazon also has a partnership with HBO, so there is a vast collection of HBO shows on Amazon Prime. Amazon Prime costs $99 per year and includes the free 2 day shipping, video, music streaming, and unlimited photo storage.

Amazon offers add-ons such as HBO, Showtime, Starz, and 30 additional channels. If you already have Amazon Prime, then this is a no brainer. If you do not, it may be something to consider, especially if you are a frequent shopper on Amazon.

Chapter 19: Vudu

Vudu is another pay streaming service owned by Walmart. Unlike the other pay streaming services, Vudu is not subscription based.

Rather, you rent or buy the individual movies and shows that you want to watch. Vudu has a catalog of over 18000 movies and 5000 television shows, and has contracts with all major movies studios and over 50 independent studios. As an added bonus, Vudu offers a few movies and shows free each month. One advantage of Vudu over the other pay services, is the ability to watch new releases before they are available on other services. They also routinely offer discounted movies and shows to rent. Vudu is available on most streaming devices, including Roku, Chromecast, Playstation, Xbox, Android and iOS. But, it does not currently support Apple TV.

Free (ad-supported) Streaming Services

Chapter 20: Tubi TV

Tubi TV is a free ad supported streaming service with thousands of movies and television shows. Content partners include Paramount Pictures, Metro-Goldwyn-Mayer (MGM), and Lionsgate, as well as over 200 others. The catalog now includes over 50,000 titles. Tubi TV is available on Android and iOS phones and mobile devices, as well as streaming devices such as Roku, Apple TV, Amazon Fire, and Chromecast. Tubi TV can also be streamed on their website. If you are looking for free movies and television, Tubi TV is a great option

Chapter 21: Crackle

Another Free ad supported streaming service is Sony's Crackle. Crackle was founded as Grouper, then purchased by Sony in 2006 and rebranded as Crackle in 2007. Crackle features many Columbia Pictures, Screen Gems, TriStar Pictures, and Sony Pictures Classics titles and Sony-distributed television series like The Shield, Seinfeld, Damages, and Rescue Me. Titles are swapped out every month, so content is always changing. This is another great free option for cord cutters and is available on most devices including Roku, Apple TV, Amazon Fire, and Chromecast. There are also Android, iOS, and Windows Phone applications for Crackle. You can also watch Crackle via any web browser.

Chapter 22: Ameba TV

Ameba TV is a free ad supported streaming service exclusively for children. Founded in 2007, Ameba TV delivers educational TV shows, cartoons, and music videos for children. Ameba TV has hundreds of different shows to choose from and includes classics such as Gumby and Popeye. There is an upgrade option to remove ads which costs $3.99 per month. Ameba TV is available on Roku, Tivo, Apple TV, and Chromecast. If you have children or grandchildren, this is a good option for free cartoons and kid shows. Another great feature of Ameba TV is they offer programming in Spanish, French, and sign language.

Streaming Devices

If you chose to use a streaming service, you need a device or smart television to watch the service via the Internet on your television. You can also watch most services on a computer, tablet, or smart phone. This section provides a short description of numerous streaming devices. However, smart televisions will not be included because of the many different versions and capabilities. Two exceptions are the Roku and Amazon Fire TVs. These both function the same as a stand alone Roku or Amazon Fire, so they are discussed. For other televisions, just be sure it has built in applications for the services you plan to watch. Also, tablets and smart phones use applications for the different streaming services, while computers use a browser. So, there is no need to discuss them here. Simply download the application or use a browser.

Chapter 23: Roku

The most popular streaming device is the Roku. Released in 2008, Roku was the first streaming device capable to stream Netflix content. While success was not overnight, Roku steadily built a loyal customer base and has continued to grow. Roku has the most channels or apps of any streaming device and you can even create your own private channels for a Roku. The openness of Roku programming allows any streaming service to develop a channel and Roku does not create their own content, so there is no bias towards any channel or content. As of August 2017, there are over 4500 channels available for the Roku and these provide access to over 450,000 movies and TV shows.

In addition to watching shows, the Roku also has a game store and is a local media player. So, you can view your own videos and pictures on your television. Roku also has the ability to mirror your cell phone screen.

The universal search feature on the Roku allows you to search across all your channels by title or actor. Outside of the Amazon Fire TV and Tivo, the Roku is the only device that gives you direct access to all of Amazon's content.

Roku has partnered with TV manufactures and allows them to build Roku TVs. So, the Roku streaming device is built into the TV and there is no need for a separate device. This is great for cord cutters because the TVs have built in tuners and it is very easy to switch between live local television and Roku channels.

Roku has 6 different versions for sale. They range in price from $29.99 for the Roku Express, to $109.99 for the Roku Ultimate. All 6 versions provide access to the same applications and channels. But, the higher end units have more options such as headphone jacks on the remotes, 4k video output, and faster processors.

Roku is a great option for cord cutting. The thousands of channels and no bias make it an excellent choice. If you are an iPhone user and purchase a lot of movies and television via iTunes, you may want to look at an Apple TV as well. Many streaming service providers have specials that include a free Roku. So, if you are interested in trying one out, look for a deal with a free device.

Chapter 24: Chromecast

The Chromecast was developed by Google and is different from other streaming devices in that you need a second device to use the

Chromecast. The Chromecast allows you to "Cast" your media from your Android or iOS phone, Android tablet or iPad, or computer, to your television via the Chromecast. The system works on your WiFi network and is easy to setup. You connect the Chromecast to your television, then download the Google Home application. The Google Home application will see the Chromecast on the same WiFi network and walk you through setting up the Chromecast. After the setup in complete, you can "cast" application media, and browser content to your television with the cast icon within the apps and browser.

The Chromecast only costs $35 and there is a 4k resolution version for $69. There are thousands of applications that support Chromecast including music, videos, games, sports, and photos. Most major streaming applications such as Netflix, Sling, Vudu, HBO Go, Crackle, Hulu, MLB TV, and Plex are included. The only downside is the requirement to have a second device to stream your content. If you always have a phone with you, that is not such a big deal, but I prefer a standalone device for my primary television. The Chromecast is a good option for a secondary TV.

Chapter 25: Apple TV

Apple TV is a top competitor to the Roku and if you are an Apple user, it may be a great alternative. Very similar to the Roku, the Apple TV is a stand alone device that has over 1600 video applications to choose from. You can also use AirPlay to stream content from your apple devices, similar to the Chromecast. One great advantage of the Apple TV is the continuous service across your Apple devices. So if you are watching a movie on your iPhone or iPad, you can pause it, then continue watching it on the Apple TV. The content is synced across your devices and knows where you stopped watching.

Apple TV also works with smart home accessories and allows you to control your Nest thermostat, or Hue lights. Apple TV also answers questions about shows, such as "what college basketball games are on today?" So, it works similar to a Google Home or Amazon Echo for your television. You can also use your iPhone or iPad as a remote control, if you misplace the included remote control.

The current version of Apple TV sells for $149.

Chapter 26: Amazon Fire TV
Another great streaming device is the Amazon Fire TV. With over 15,000 applications and games available, the Amazon Fire TV allows you to watch almost any streaming service. As might be expected, the Amazon Fire TV is centered around Amazon content, but it does support most other streaming services as well.

There are two options for the Fire TV, the Stick or the box. The First TV stick costs $39.99 and the Fire TV costs $89.99. If you plan to use a lot of the games, the Fire TV is a better option, otherwise the Fire Stick can do everything you need. The Fire TV Stick is so popular; it is often out of stock on Amazon. Both products include the universal search function for over 140 channels, and the voice search function via the remote control. You may remember the comical Gary Busey commercials for the voice search. But it does work very well, and will do more than just search. You can say commands such as, "watch Game of Thrones" or "launch Netflix." It is based on the Amazon Echo Alexa and can do pretty much everything an Amazon Echo can.

If you are a heavy Amazon user, the Fire TV is a great option at a great price.

Section 3

Chapter 27: Free Television
With the increasing cost of cable and satellite television many people consider cord cutting to reduce costs. But, what about eliminating it all together. Outside of equipment costs the following services have no costs and will provide you a wealth of entertainment options.

The first option is the over the air (OTA) broadcasts. If you can receive broadcast transmissions at your residence, this is a great option for free television. To supplement OTA television there are multiple free Internet streaming options. Of course you must have Internet access and a streaming device (phone, tablet, computer, smart TV, Roku, etc) to watch the free Internet options. As described in the previous sections you can watch Pluto TV, Crackle, Tubi TV, and Ameba TV for free. Many networks also provide limited free content, as do cable channels through their websites or apps. With these options, you have access to thousands of movies, TV shows, and live television. You really can do this for free!

Chapter 28: Sports

If you are sports fanatic, there are multiple options for watching your favorite sports and teams. 3 of the 4 major sports have live streaming options.

Major League Baseball - MLB.TV

National Basketball League – NBA League Pass and NBA TV

National Hockey League – NHL.TV

For the National Football League there are few options. The first is NFL Sunday Ticket which is available as a streaming option if you are in college or live in an apartment or condo. If that doesn't work for you, then you can watch all the games after they are complete on NFL Game Pass. It isn't live, but it is better than not watching the game. Of course, with an antenna you will get local NFL games.

For college sports, you have the ESPN networks, Fox Sports Regional, NBC Sports Network, the Big Ten Network, and the SEC Network available with different streaming services. Review the table below to see what services offer which networks.

The Golf channel is available on all 6 pay streaming services.

Chapter 29: Movies

As you have already read, there are multiple options for watching movies. Between Netflix, Amazon Prime, Hulu, and Vudu, you can watch thousands of movies. But, you also have the free options with Crackle, Pluto TV, and Tubi TV. You will be very surprised at the quality free movies available on those services.

Chapter 30: News

For news, you have the option of local news via OTA or national news on one of the streaming services. Plus, Pluto TV has free access to NBC/MSNBC, and Sky News. Most cable news channels are available on one of the streaming services as well.

Conclusion

I hope you have enjoyed my book and it has enabled you to make an informed decision if you decide to cut the cord. The most important decision is to determine what shows or sports you must have. From there, review the different options and see how you can watch those specific shows and sports. Do not fall into the old cable trap, that you must have specific channels all the time. All of these cord cutting options are contract free. So, if there is one show you must watch live, you have the option to pay for a service that offers that show while it is airing. Then you can cancel the service. For example, I like the Walking Dead. So, I pay for Sling while the Walking Dead is airing live because it is the cheapest service that offers AMC. Afterwards, I cancel my Sling service. I do the same thing with HBO Now so I can watch Game of Thrones. Those are the only 2 services I pay for at different times throughout the year. Otherwise, my television is FREE!

Be sure to review the appendix for tables of streaming devices, local channels and live television via streaming services.

If you would like updated information as it becomes available, consider following my Facebook page at https://www.facebook.com/cordcuttingnews/ and on Twitter at https://twitter.com/cordcutnews.

Appendix

Streaming Device and Live Television Services

Streaming Player	Sling TV	PS Vue	Direct TV Now	Youtube TV	Hulu Live TV	Fubo TV	Pluto TV
Apple TV	X	X	X	X	X	X	X
Amazon Fire TV	X	X	X		X	X	X
Roku	X	X	X			X	X
Chromecast	X	X	X	X	X	X	X
Android TV	X	X				X	X
Xiaomi	X						
Leeco	X						
AirTV	X						
ZTE	X						
Channel Master	X						
LG WebOS	X						
IOS iPhones	X	X	X	X	X	X	X
Android Phones	X	X	X	X	X	X	X
Amazon Fire Phones	X						
Xbox One	X				X		
OSX (apple)	X	X	X	X		X	X
Windows 10	X	X	X	X		X	X
PlayStation 4		X					
DVR	Yes	Yes	No	Yes	Yes	Yes	No
Cost	$20+	$40+	$35+	$35+	$40+	$35+	Free

Live Television Channels available on Streaming Services

Channel	Sling TV	PS Vue	DIRECTV NOW	FuboTV Premier	Hulu Live TV	Youtube TV
A&E	X		X	X	X	
AMC	X	X	X			X
American Heroes Channel		X	X			
Animal Planet		X	X			
Audience			X			
Baby TV	X					
BBC America	X	X	X			X
BBC World News	X	X	X			
Bein Sports				X		
Blaze TV	X					
BET	X		X			
Big Ten Network		X	X	X	X	X
Bloomberg Television	X					
Boomerang	X	X			X	
Bravo	X	X	X	X		X
Campus Insider	X					
Cartoon Network/Adult	X	X	X		X	

Swim							
CBS Sports Network						X	X
Centric			X				
Chiller		X	X	X	X		X
Cinemax	X	X	X				
Cheddar	X						
Cloo		X	X				
CMT	X		X				
Comedy TV			X				
CNBC	X	X	X	X		X	X
CNBC World		X	X	X			
CNN	X	X	X			X	
CNN International						X	
AXS TV	X						
Comcast Sportsnet	X	X	X				X
Comedy Central	X		X				
Crime & Investigatin			X				
Cooking Channel	X	X	X				
Destination America		X	X				

Discovery		X	X			
Discovery Family		X	X			
Discovery Life		X	X			
Disney Channel	X	X	X		X	X
Disney Junior	X	X	X		X	X
Disney XD	X	X	X		X	X
DIY Network	X	X	X			
Duck TV	X					
E!	X	X	X	X	X	X
Eleven Sports				X		
EPIX	X					
EPIX Drive-In	X					
EPIX Hits	X	X				
EPIX2	X					
ESPN	X	X	X		X	X
ESPN 2	X	X	X		X	X
ESPN NEWS					X	
ESPN Bases Loaded	X					
ESPN Buzzer Beater	X					
ESPN Deportes	X	X				
ESPN Goal Line	X	X				

ESPN U	X	X	X		X	X
ESPN News	X	X	X			X
El Rey	X		X	X		
Esquire		X				
Euronews	X					
FM			X			
Fla Ma	X					
Food Network	X	X	X		X	
Football Report TV				X		
Fox Business		X	X		X	X
Fox College Sports		X				
Fox News Channel		X	X	X	X	X
Fox Regional Sports Networks	X	X	X	X		
Fox Soccer Plus				X		X
Fox Sports						X
France 24	X					
FreeForm	X	X	X		X	X
FS1	X	X	X	X	X	X
FS2	X	X	X	X	X	X
Fuse				X		

FX	X	X	X	X	X	X	
FXM		X	X	X	X	X	
FXX	X	X	X	X	X	X	
FYI	X		X	X	X		
Galavision				X			
Golf Channel	X	X	X	X	X	X	
Great America			X				
Hallmark Channel	X		X	X			
Hallmark Movies and Mysteries				X			
HBO	X	X	X				
HDNet Movies	X						
HGTV	X	X	X		X		
History	X		X	X	X		
Hi-Yah!		X					
HLN	X	X	X		X		
GSN	X		X				
IFC	X	X	X			X	
Investigation Discovery		X	X				
Lifetime	X		X	X	X		
Lifetime Movies	X		X	X	X		

Local Now	X			X			
Logo	X		X				
Machinima		X					
Maker	X						
MGM		X					
MSNBC	X	X	X		X		X
MTV	X		X	X			
MTV 2	X		X				
Nat Geo Wild	X	X	X	X	X		X
Nat Geo	X	X	X	X	X		X
NBA TV	X	X	X	X			
NBC Sports Network	X	X	X		X		X
NFL Network	X	X					
NDTV	X						
NHL Network	X		X				
News18	X						
Newsy	X	X					
Nickelodeon			X				
Nick Jr.	X		X				
Nicktoons	X		X				
One World Sports		X					
Outside TV	X	X					

OWN		X	X				
Oxygen	X	X	X	X	X		X
Pursuit Channel			X				
Pac-12	X						
Poker		X					
Polaris+	X	X					
Pop		X			X		
RT Network	X						
Science Channel		X	X				
SEC Network	X	X	X		X		X
Showtime		X					X
Sony Movie Channel		X					
Spike	X		X				
Sprout		X	X	X	X		X
Sundance TV	X	X	X				X
SyFy	X	X	X	X	X		X
TBS	X	X	X		X		
TeenNick	X		X				
Telemundo							X
TLC		X	X				
TNT	X	X	X		X		
Travel Channel	X	X	X		X		

Tru TV	X	X	X		X	
Turner Classic Movies	X	X	X		X	
TV Land	X		X			
Universal		X		X		X
Unimas	X		X	X		
Universo (NBC)						X
Univision	X		X	X		
USA	X	X	X	X	X	X
Velocity		X	X			
VH1	X		X			
Vibrant	X					
Viceland	X		X	X	X	
WE tv	X	X				X
Fox Deportes	X	X				
Weather Channel			X	X		
WeatherNation			X			

Live Local Channels
DirecTV Now, Hulu Live

	DirecTV Now			Hulu with Live TV			
Network channel:	abc	fox	nbc	abc	cbs	fox	nbc
Totals by channel:	39	39	15	8	119	50	11
Total for each service:		93			188		
TOP 50 MARKETS							
New York	abc	fox	nbc	abc	cbs	fox	nbc
Los Angeles	abc	fox	nbc	abc	cbs	fox	nbc
Chicago	abc	fox	nbc	abc	cbs	fox	nbc
Philadelphia	abc	fox	nbc	abc	cbs	fox	nbc
Dallas-Ft. Worth	abc	fox	nbc		cbs	fox	nbc
San Francisco-Oakland	abc	fox	nbc	abc	cbs	fox	nbc
Washington DC	abc	fox	nbc		cbs	fox	nbc
Houston	abc	fox		abc	cbs	fox	
Boston	abc		nbc		cbs		nbc
Atlanta	abc	fox			cbs	fox	
Tampa-St Petersburg	abc	fox			cbs	fox	
Phoenix	abc	fox			cbs	fox	
Detroit	abc	fox			cbs	fox	
Seattle-Tacoma	abc	fox			cbs	fox	
Minneapolis-St. Paul	abc	fox			cbs	fox	
Miami-Ft. Lauderdale	abc	fox	nbc		cbs	fox	nbc
Denver	abc	fox			cbs	fox	
Orlando-Daytona Beach	abc	fox			cbs	fox	
Cleveland-Akron	abc	fox			cbs	fox	
Sacramento-Stockton	abc	fox			cbs	fox	
St. Louis	abc	fox			cbs	fox	
Charlotte	abc	fox			cbs	fox	
Pittsburgh	abc		nbc		cbs		
Raleigh-Durham	abc			abc			
Portland (OR)	abc					fox	
Baltimore	abc				cbs		
Indianapolis	abc	fox				fox	
San Diego	abc	fox	nbc		cbs	fox	nbc
Nashville					cbs		
Hartford-New Haven	abc	fox			cbs	fox	nbc
San Antonio	abc				cbs		
Columbus (OH)	abc				cbs		

Market	abc	fox	nbc		cbs	fox	
Kansas City	abc	fox	nbc		cbs	fox	
Salt Lake City		fox	nbc			fox	
Milwaukee	abc	fox	nbc		cbs	fox	
Cincinnati	abc						
Greenville-Spartanburg	abc						
West Palm Beach	abc		nbc				
Austin	abc	fox				fox	
Las Vegas (NV)	abc					fox	
Oklahoma City					cbs		
Norfolk-Portsmouth							
Harrisburg-Lancaster (PA)		fox				fox	
Grand Rapids-Kalamazoo		fox				fox	
Birmingham, AL							
Greensboro, Winston-Salem		fox			cbs	fox	
Jacksonville					cbs		
Louisville					cbs	fox	
New Orleans					cbs	fox	
OTHER MARKETS*	*Only markets with at least one local affiliate are listed.						
Albany (GA)					cbs		
Albany (NY)							
Alexandria (LA)					cbs		
Amarillo					cbs		
Anchorage (AK)		fox				fox	
Augusta-Aiken					cbs		
Bangor		fox			cbs		
Baton Rouge					cbs		
Bend (OR)					cbs	fox	
Biloxi-Gulfport		fox			cbs	fox	
Binghamton (NY)					cbs		
Bowling Green (KY)					cbs		
Burlington-Plattsburgh					cbs		
Charleston (SC)					cbs		
Charlottesville					cbs		
Chattanooga					cbs		
Cheyenne-Scottsbluff					cbs		
Chico-Redding					cbs		
Clarksburg-Weston					cbs		
Colorado Springs-Pueblo					cbs		

Columbia (SC)					cbs	
Columbia-Jefferson City						fox
Columbus-Tupelo(MS)		fox			cbs	fox
Dayton					cbs	
Des Moines-Ames					cbs	
Dothan					cbs	
Duluth-Superior					cbs	
Elmira (Corning)					cbs	
Erie					cbs	
Eugene		fox				
Eureka (CA)					cbs	
Evansville (IN)					cbs	
Fairbanks (AK)					cbs	
Fargo-Valley City					cbs	
Flint-Saginaw-Bay City					cbs	
Fresno-Visalia	abc			abc		
Ft. Myers-Naples					cbs	
Gainesville		fox				fox
Greenwood-Greenville (MS)					cbs	
Brownsville-McAllen						
Harrisonburg (VA)					cbs	
Hattiesburg/Laurel (MS)						fox
Honolulu					cbs	
Idaho Falls-Jackson					cbs	fox
Jackson (TN)					cbs	
Jonesboro (AR)						fox
Juneau (AK)		fox				fox
Knoxville					cbs	
La Crosse-Eau Claire					cbs	
Lafayette (IN)						fox
Laredo					cbs	
Lexington					cbs	
Lima (OH)					cbs	
Lincoln-Hastings-Kearney					cbs	
Little Rock-Pine Bluff					cbs	
Macon (GA)					cbs	
Madison (WI)					cbs	
Mankato (MN)					cbs	
Meridian (MS)					cbs	
Monroe-El Dorado					cbs	

Monterey-Salinas					cbs		
Montgomery-Selma					cbs		
North Platte					cbs		
Odessa-Midland					cbs		
Omaha (NE)					cbs		
Paducah-Cape Girardeau					cbs		
Palm Springs					cbs	fox	
Panama City					cbs		
Parkersburg					cbs		
Presque Isle (ME)					cbs		
Reno (NV)					cbs		
Roanoke-Lynchburg					cbs		
Rockford (IL)					cbs		
Salisbury (MD)					cbs		
Santa Barbara-Santa Maria					cbs	fox	
Savannah					cbs		
Sherman-Ada					cbs		
Shreveport					cbs		
Spokane					cbs		
Springfield (MO)			fox				
Springfield-Holyoke (MA)					cbs		
St. Joseph						fox	
Tallahassee-Thomasville					cbs		
Toledo					cbs		
Topeka					cbs		
Traverse City-Sault Ste.Marie					cbs		
Tucson (Sierra Vista)					cbs		
Tulsa					cbs		
Twin Falls					cbs		
Tyler-Longview					cbs		
Waco-Temple-Bryan					cbs		
Watertown (NY)					cbs		
Wausau-Rhinelander					cbs		
Wichita Falls & Lawton					cbs		
Wichita-Hutchinson					cbs		
Wilmington, NC					cbs		
Yuma-El Centro						fox	

Sling TV, YouTube TV

	Sling TV			YouTube TV			
Network channel:	abc	fox	nbc	abc	cbs	fox	nbc
Totals by channel:	8	17	11	24	28	26	27
Total for each service:	36			105			
TOP 50 MARKETS							
New York	abc	fox	nbc	abc	cbs	fox	nbc
Los Angeles	abc	fox	nbc	abc	cbs	fox	nbc
Chicago	abc	fox	nbc	abc	cbs	fox	nbc
Philadelphia	abc	fox	nbc	abc	cbs	fox	nbc
Dallas-Ft. Worth		fox	nbc		cbs	fox	nbc
San Francisco-Oakland	abc	fox	nbc	abc	cbs	fox	nbc
Washington DC		fox	nbc	abc	cbs	fox	nbc
Houston	abc	fox		abc	cbs	fox	nbc
Boston			nbc	abc	cbs	fox	nbc
Atlanta		fox		abc	cbs	fox	nbc
Tampa-St Petersburg		fox		abc	cbs	fox	
Phoenix		fox		abc	cbs	fox	nbc
Detroit		fox		abc	cbs	fox	nbc
Seattle-Tacoma				abc	cbs		nbc
Minneapolis-St. Paul		fox		abc	cbs	fox	nbc
Miami-Ft. Lauderdale			nbc	abc	cbs	fox	nbc
Denver							
Orlando-Daytona Beach		fox		abc	cbs	fox	nbc
Cleveland-Akron							
Sacramento-Stockton							
St. Louis							
Charlotte		fox		abc	cbs	fox	nbc
Pittsburgh				abc	cbs	fox	nbc
Raleigh-Durham	abc						
Portland (OR)							
Baltimore				abc	cbs	fox	nbc
Indianapolis							
San Diego			nbc				
Nashville					cbs	fox	nbc
Hartford-New Haven			nbc				
San Antonio				abc	cbs	fox	nbc
Columbus (OH)				abc	cbs	fox	
Kansas City							
Salt Lake City							
Milwaukee							

Market				abc	cbs		fox	nbc
Cincinnati				abc	cbs			nbc
Greenville-Spartanburg								
West Palm Beach				abc	cbs			nbc
Austin			fox					
Las Vegas (NV)				abc			fox	nbc
Oklahoma City								
Norfolk-Portsmouth								
Harrisburg-Lancaster (PA)								
Grand Rapids-Kalamazoo								
Birmingham, AL								
Greensboro, Winston-Salem								
Jacksonville						cbs	fox	nbc
Albuquerque-Santa Fe								
Louisville						cbs	fox	nbc
New Orleans								
OTHER MARKETS*								
Fresno-Visalia		abc						
Ft. Myers-Naples								
Gainesville			fox					
Memphis						cbs	fox	nbc

PlayStation Vue

	PlayStation Vue			
Network channel:	abc	cbs	fox	nbc
Totals by channel:	19	104	35	17
Total for each service:	175			
TOP 50 MARKETS				
New York	abc	cbs	fox	nbc
Los Angeles	abc	cbs	fox	nbc
Chicago	abc	cbs	fox	nbc
Philadelphia	abc	cbs	fox	nbc
Dallas-Ft. Worth	abc	cbs	fox	nbc
San Francisco-Oakland	abc	cbs	fox	nbc
Washington DC		cbs	fox	nbc
Houston	abc	cbs	fox	nbc
Boston		cbs		nbc

Atlanta	abc	cbs	fox	
Tampa-St Petersburg	abc	cbs	fox	
Phoenix	abc	cbs	fox	
Detroit	abc	cbs	fox	nbc
Seattle-Tacoma		cbs	fox	
Minneapolis-St. Paul		cbs	fox	
Miami-Ft. Lauderdale		cbs	fox	nbc
Denver	abc	cbs	fox	
Orlando-Daytona Beach	abc	cbs	fox	nbc
Cleveland-Akron	abc	cbs	fox	
Sacramento-Stockton	abc	cbs	fox	nbc
St. Louis		cbs	fox	
Charlotte	abc	cbs	fox	
Pittsburgh	abc	cbs	fox	nbc
Raleigh-Durham	abc			
Portland (OR)			fox	
Baltimore		cbs	fox	nbc
Indianapolis			fox	
San Diego		cbs	fox	nbc
Nashville			fox	
Hartford-New Haven		cbs	fox	nbc
San Antonio		cbs	fox	
Columbus (OH)		cbs	fox	
Kansas City		cbs	fox	
Salt Lake City		cbs	fox	
Milwaukee		cbs	fox	
Cincinnati		cbs		
Greenville-Spartanburg				
West Palm Beach		cbs		
Austin		cbs	fox	
Las Vegas (NV)				
Oklahoma City		cbs		
Norfolk-Portsmouth				
Harrisburg-Lancaster (PA)		cbs		
Grand Rapids-Kalamazoo		cbs		
Birmingham, AL				
Greensboro, Winston-Salem		cbs		
Jacksonville		cbs		
Albuquerque-Santa Fe				
Louisville		cbs		

New Orleans		cbs		
OTHER MARKETS*				
Albany (GA)		cbs		
Albany (NY)		cbs		
Alexandria (LA)		cbs		
Amarillo		cbs		
Augusta-Aiken		cbs		
Baton Rouge		cbs		
Biloxi-Gulfport		cbs		
Burlington-Plattsburgh		cbs		
Cedar Rapids-Iowa City		cbs		
Charleston (SC)		cbs		
Charlottesville		cbs		
Cheyenne-Scottsbluff		cbs		
Colorado Springs-Pueblo		cbs		
Dayton		cbs		
Des Moines-Ames		cbs		
Dothan		cbs		
El Paso (Las Cruces)		cbs		
Elmira (Corning)		cbs		
Erie		cbs		
Fargo-Valley City		cbs		
Flint-Saginaw-Bay City		cbs		
Fresno-Visalia	abc			
Ft. Myers-Naples		cbs		
Gainesville			fox	
Brownsville-McAllen		cbs		
Honolulu		cbs		
Idaho Falls-Jackson		cbs		
Jackson (TN)		cbs		
Knoxville		cbs		
Laredo		cbs		
Lexington		cbs		
Lincoln-Hastings-Kearney		cbs		
Meridian (MS)		cbs		
Monroe-El Dorado		cbs		
Monterey-Salinas		cbs		
Montgomery-Selma		cbs		
North Platte		cbs		
Odessa-Midland		cbs		

Paducah-Cape Girardeau		cbs		
Palm Springs		cbs		
Panama City		cbs		
Parkersburg		cbs		
Portland-Auburn (ME)		cbs		
Presque Isle (ME)		cbs		
Reno (NV)		cbs		
Roanoke-Lynchburg		cbs		
Rockford (IL)		cbs		
Salisbury (MD)		cbs		
Santa Barbara-Santa Maria		cbs		
Savannah		cbs		
Sherman-Ada		cbs		
Shreveport		cbs		
South Bend-Elkhart		cbs		
Springfield-Holyoke (MA)		cbs		
Tallahassee-Thomasville		cbs		
Toledo		cbs		
Topeka		cbs		
Traverse City-Sault Ste.Marie		cbs		
Tucson (Sierra Vista)		cbs		
Tulsa		cbs		
Twin Falls		cbs		
Waco-Temple-Bryan		cbs		
Wausau-Rhinelander		cbs		
Wichita Falls & Lawton		cbs		
Wichita-Hutchinson		cbs		

ABOUT THE AUTHOR

Dr. Thomas Hyslip is currently the Resident Agent in Charge of the Department of Defense, Defense Criminal Investigative Service (DCIS), Cyber Field Office, Eastern Resident Agency. Prior to joining the DCIS in 2007, Dr. Hyslip was a Special Agent with the US Environmental Protection Agency, Criminal Investigation Division, and the US Secret Service. Throughout his 17 years of federal law enforcement, Dr. Hyslip has specialized in cybercrime investigations and computer forensics. Dr. Hyslip has testified as an expert witness on computer forensics and network intrusions at numerous federal, state, and local courts.

Dr. Hyslip is also a Colonel in the U.S. Army Reserves and is currently assigned as an Assistant Professor of Preventive Medicine at the F. Edward Hébert School of Medicine, Uniformed Services University of the Health Sciences. COL Hyslip has a mix of active duty and reserve assignments spanning over 20 years including assignments with the US African Command, Office of Inspector General, the Department of Defense, Office of Inspector General, and the US Army Reserve Information Operations Command. In 2005 COL Hyslip deployed to Iraq with the 306th Military Police Battalion and earned a Bronze Star, Purple Heart, and Combat Action Badge.

Dr. Hyslip is currently an adjunct Professor at Norwich University Dr. Hyslip received his Doctor of Science degree in Information Assurance from Capitol College in 2014. Dr. Hyslip previously obtained a Master of Science degree from East Carolina University and a Bachelor of Science degree from Clarkson University.

Made in United States
Orlando, FL
09 February 2024